How to Study Calculus

Joseph Mazur

WCB Wm. C. Brown Publishers
Dubuque, Iowa•Melbourne, Australia•Oxford, England

Wm. C. Brown Communications, Inc.

Copyright © 1994 by Wm. C. Brown Communications, Inc. All rights reserved

Cover design by Jeanne Calabrese

A Times Mirror Company

Library of Congress Catalog Card Number: 93-70439

ISBN 0-697-20197-X

No part of this publication may be reproduced, stored in a retrieval system, or transmitted, in any form or by any means, electronic, mechanical, photocopying, recording, or otherwise, without the prior written permission of the publisher.

Printed in the United States of America by Wm. C. Brown Communications, Inc., 2460 Kerper Boulevard, Dubuque, IA 52001

10 9 8 7 6 5 4

Contents

Introduction		*1*
Diagnostic Evaluation		*2*

1 Scheduling and Organizing   4

    Creating a work environment   5
    Assessing how much time you will need   6
    Falling behind and working ahead   6
    Priming your studies   7

2 Listening and Participating   8

    Attendance   9
    Class participation   9
    How to listen   10
    What to do when you get lost   11
    Recalling the previous lecture   12
    Coordinating your reading with your listening   13

3 Taking Notes   14

    What are the important points?   15
    How much to write   16
    How to use what you have written   17
    Sample notes   17

4 Reading the Text   19

    How to begin   20
    Using the margins   21
    Highlighting and how to use it   21
    Organizing your reading   22

5 Working through a Problem — 23

- Organizing your thoughts — 24
- Isolating the question — 25
- Developing strategies — 25
- Simplifying the problem — 26
- Breaking down the problem — 26
- Organizing your worksheet — 27
- Interpreting your results — 28
- Develop good habits — 29
- Work with classmates — 29
- Use your instructor — 29

6 Preparation of Assignments — 30

- Clarification of what is expected — 31
- Presentation of your work — 31
- Page layout — 31
- Drawing useful diagrams and graphs — 32
- Displaying the answers — 33

7 Preparing for quizzes and taking exams — 34

- Preparation — 35
- Using your notes — 35
- Using your text — 36
- Using returned assignments — 36
- Using computers and video tutorials — 37
- Assessing your strengths and weaknesses — 37
- Getting help — 37
- Friendly advice — 38

8 At the Exam — 39

- Relax — 40
- How to take an exam — 40

9 Summary — 42

Introduction

Calculus is the entrance course for almost all mathematics, engineering and science courses. Each year, more than a half-million students enter calculus courses in the United States and more than 175,000 students fail to pass. This book is about what you can do to increase your own efficiency in learning the subject. Learning calculus requires your active participation; it requires an efficiency in learning. Developing that efficiency is part of the process. This book is designed to assist you. But it is up to you. Good luck.

Why this book?

Motivation, good teaching and aptitude are important for successful learning of a subject; but without efficiency, your studies can be slow, confusing and fatiguing. I have been teaching calculus for almost twenty years and have observed well motivated students with high aptitude receive grades that do not reflect their abilities. Many students miss opportunities for better grades simply because they did not know how to study, listen, take notes or prepare for an exam. The study skills for learning calculus are slightly more special than those for history or social studies. This book should provide you with a faster way to understand those special skills.

How to use this book

The order in which the topics appear is unimportant. While all topics are useful in exercising your efficiency, it is more important to commit yourself to the development of each skill discussed, than simply to read through the material.

The book is designed for a quick reading (you have enough to read in your classes); but it will take thought and work to develop the habits suggested in these topics. Even so, the extra work will, in the long run, save time and energy in your calculus studies. And besides, many of the skills that you will develop from these exercises will be transferable to other courses. It is easier to write about how to study calculus than to practice the ideas that are outlined in this book; but, if you begin to practice just a few, you will have profited from the reading.

The book is divided into 8 sections

1. Time management
2. Listening to lectures
3. Taking notes
4. Reading the text
5. Solving problems
6. Preparing Assignments
7. Preparing for exams
8. Taking exams

Each section is introduced by a questionnaire that evaluates your progress throughout the term. You want to honestly answer "yes" to as many questions as possible. There are 12 questions for each section. You should strive toward having at least 9 of your answers positive in each section. Re-evaluate your skills each week, until you feel that you have comfortably improved in implementing the skills that are talked about in each section.

There are 50 tips listed in this book. They are listed at the end of each section and are placed there as summaries of what you have read. Review these tips from time to time to remind yourself to absorb them into your study habits.

The Diagnostic Evaluation

The diagnostic evaluation that follows on the next page is designed to assess your background in algebra and to partially determine your readiness for calculus. All problems represent the algebraic and trigonometric information that you must know very well in order to solve the problems in the first six chapters of standard calculus texts.

The evaluation recognizes that your algebra skills will improve as you move through the course but (at the same time) indicates the areas of your weaknesses so that you may improve your skills in such areas.

This evaluation is based on an assessment of the bare minimum background required for calculus. Most instructors will expect much more. If you have difficulty with any one of the evaluation problems, you should consult your advisor or instructor about his or her expectations in the course.

Diagnostic Evaluation

1) $f(x) = x^{-1} + x^0$. Find $f(x+2)$ and simplify.

2) Solve the following equation for h: $x^2 + 4xh = 108$.

3) Solve for x: a) $x^2 - 14 = 2$ b) $\dfrac{3x^2 + 1}{x - 1} = 2x$

4) Solve for x: a) $2^{x^2} = 4^x$ b) $2^{2x-1} = 13$

5) In the figure below, the curved path is made up of 8 semicircles of equal diameter. If the total length of the curved path is 16π units, then find the area of the rectangle.

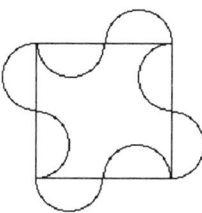

6) Solve the following equation for x: $1 + \sqrt{x} = 2x$

7) Solve simultaneously: $2x + y = 8;\ x^2 - 7y = 9$

8) Solve simultaneously $\dfrac{4}{x} + \dfrac{3}{y} = 3$ and $\dfrac{12}{y} - \dfrac{2}{x} = 3$

9) Sketch the graphs of: a) $x = 3$ b) $y = x^2 + 4$ c) $y = (x - 1)^2$

10) Sketch the graphs of: a) $y = -4$ b) $y = 4 - x^2$ c) $x = (y - 4)^2 - 2$

11) Simplify the ratio: $\dfrac{1}{1 - \dfrac{1}{1 + x}}$

12) Given a right triangle with $\omega = 30$ degrees and $y = 3$, what are the values for x and z?

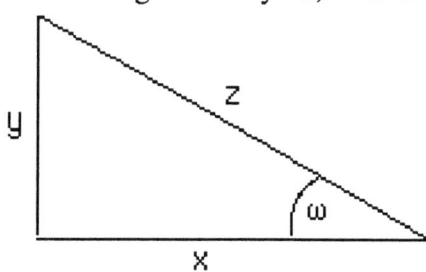

Scheduling and Organizing 1

How Organized are you?

	Now	week 1	week 2	week 3	week 4
Do you ...	↓	↓	↓	↓	↓
1. have a comfortable place to do your work? _____	☐ *Yes* ☐ *No*	☐ *Yes* ☐ *No*	☐ *Yes* ☐ *No*	☐ *Yes* ☐ *No*	☐ *Yes* ☐ *No*
2. do your writing at a table? _____	☐ *Yes* ☐ *No*	☐ *Yes* ☐ *No*	☐ *Yes* ☐ *No*	☐ *Yes* ☐ *No*	☐ *Yes* ☐ *No*
3. have adequate lighting at your work area? _____	☐ *Yes* ☐ *No*	☐ *Yes* ☐ *No*	☐ *Yes* ☐ *No*	☐ *Yes* ☐ *No*	☐ *Yes* ☐ *No*
4. do your assignments in a quiet place? _____	☐ *Yes* ☐ *No*	☐ *Yes* ☐ *No*	☐ *Yes* ☐ *No*	☐ *Yes* ☐ *No*	☐ *Yes* ☐ *No*
5. keep a schedule of your daily routines? _____	☐ *Yes* ☐ *No*	☐ *Yes* ☐ *No*	☐ *Yes* ☐ *No*	☐ *Yes* ☐ *No*	☐ *Yes* ☐ *No*
6. keep a schedule of your long term deadlines? _____	☐ *Yes* ☐ *No*	☐ *Yes* ☐ *No*	☐ *Yes* ☐ *No*	☐ *Yes* ☐ *No*	☐ *Yes* ☐ *No*
7. finish your daily assignments on time? _____	☐ *Yes* ☐ *No*	☐ *Yes* ☐ *No*	☐ *Yes* ☐ *No*	☐ *Yes* ☐ *No*	☐ *Yes* ☐ *No*
8. finish your long term assignments on time? _____	☐ *Yes* ☐ *No*	☐ *Yes* ☐ *No*	☐ *Yes* ☐ *No*	☐ *Yes* ☐ *No*	☐ *Yes* ☐ *No*

9. discourage interruptions while you are at work?	☐ <u>Yes</u> ☐ <u>No</u>	☐ <u>Yes</u> ☐ <u>No</u>	☐ <u>Yes</u> ☐ <u>No</u>	☐ <u>Yes</u> ☐ <u>No</u>	☐ <u>Yes</u> ☐ <u>No</u>
10. take occasional breaks but not too often?	☐ <u>Yes</u> ☐ <u>No</u>	☐ <u>Yes</u> ☐ <u>No</u>	☐ <u>Yes</u> ☐ <u>No</u>	☐ <u>Yes</u> ☐ <u>No</u>	☐ <u>Yes</u> ☐ <u>No</u>
11. prevent yourself from daydreaming?	☐ <u>Yes</u> ☐ <u>No</u>	☐ <u>Yes</u> ☐ <u>No</u>	☐ <u>Yes</u> ☐ <u>No</u>	☐ <u>Yes</u> ☐ <u>No</u>	☐ <u>Yes</u> ☐ <u>No</u>
12. prevent yourself from procrastinating?	☐ <u>Yes</u> ☐ <u>No</u>	☐ <u>Yes</u> ☐ <u>No</u>	☐ <u>Yes</u> ☐ <u>No</u>	☐ <u>Yes</u> ☐ <u>No</u>	☐ <u>Yes</u> ☐ <u>No</u>

Creating a Work Environment

A good carpenter finishes each day by sweeping up and putting away his or her tools, knowing that the time spent in doing so will be credited to later work. Time may be spent productively and efficiently, or wasted. The time you set aside for your studies each day is likely to be short in comparison to the length of the day; so it should be used with keen frugality. How much time is wasted each day in finding an assignment, in finding your textbook or notebook, or in finding the proper pencil? How many times do you say to yourself, "I know I put that _____ on my desk. It's impossible, but it has disappeared!"

An organized work environment means organized assignments, organized thoughts and organized ideas. An organized environment helps you to use your time effectively. A two-hour study session should not have to include a twenty minute start-up time.

An organized environment means having a workspace that is clean, neat and orderly. Your desk should be cleared so that a notebook, pad and textbook may be opened without overlap. There should be spare space on the desk so that the pages may be turned without interference.

Good lighting is critical. Poor lighting can give an imperceptible eye strain, causing mental fatigue and sluggish mental activity. Avoid using lighting that causes shadows across your work area. Invest in a pencil sharpener. A fine point on a pencil adds to the neatness and clarity of your work. Throw away smudged erasers.

Assessing how much time you will need

Success in calculus requires habit and routine. Evenly divide your study time between the days of the week so that you build routine and work by habit. It is far better to work for several hours each day than to work for a random stretch of hours scattered over the term or to cram all night before a quiz.

For most colleges, the number of *credit hours* assigned to a course is related to the **minimal** number of hours per week that you are expected to prepare for that course. You should check with your advisor or student handbook to establish the correct formula for your school. And remember: The minimal number of hours is just that-the number of hours of study for only a passing grade. To do well in the subject, you may have to put in four or five hours a day.

Write out a day-to-day schedule, including a rough outline of the times for classes, other studies, dinner and social activities. Keep two schedules-a daily schedule and a long-term schedule.

The long-term schedule is as important as the daily schedule. It should signal important due dates for projects, lab reports, term papers, exams and library books due. For this, you should have an appointment book or a calendar. Even better: Get a wall calendar that displays the entire year! You will soon discover just how short the term really is.

Falling behind and working ahead

At the end of several weeks you should know if your schedule is giving you enough time to achieve your goals. You may find that you need to assign more time to one subject than another. One subject may turn out easier than you expected, while another may turn out more difficult. Just be sure that you do not shortchange yourself. The next week you may encounter a more difficult topic in the subject that you thought was easy.

The advice here is to be flexible in your scheduling; but, at the same time, give yourself ample time to keep pace with the course.

Priming your studies

Do you put off your studies? Do you sit down at your desk with every intention to study but find some excuse for doing something else instead? You sit down and find that you need a drink or a snack or you have forgotten to call someone You make yourself a cup of coffee, sit down, look at a sentence in your text and feel sleepy. You lie down.

It's okay to feel hungry and tired but when you feel that way every time you sit at your desk with your textbook, it is likely that you are missing something. You need to be primed to study well. Like a water pump that works smoothly only after a little water starts to flow, your studies will flow after you have made just a little progress.

How do you prime your studies? By starting slow. Say to yourself, "I'll work for five minutes, concentrating. I'll do at least one problem, or at least try to understand it." The reason for telling yourself that you will work for five minutes is that there is no heavy commitment. Chances are that you will loose track of the time and continue beyond the time limit you set for yourself-this is desirable. In other words, you are trying to fool yourself into believing that you will work for only five minutes; you may forget your time limitation and find yourself involved with your work for a far longer stretch than just five minutes. Had you begun with the intention of spending two hours at your studies, it is far more likely that you would have wasted half the time under the belief that you had plenty of time anyway.

Tips

- Find a quiet, comfortable place to work
- Make short- and long-term schedules
- Discourage interruptions
- Take occasional, short breaks
- Do not procrastinate
- Allow five minutes warm-up time

2

Listening and Participating

How well do you listen?

Do you ...	Now	week 1	week 2	week 3	week 4
1. attend class regularly?	☐ Yes ☐ No	☐ Yes ☐ No	☐ Yes ☐ No	☐ Yes ☐ No	☐ Yes ☐ No
2. ask questions in class or during office hours?	☐ Yes ☐ No	☐ Yes ☐ No	☐ Yes ☐ No	☐ Yes ☐ No	☐ Yes ☐ No
3. write down assignments?	☐ Yes ☐ No	☐ Yes ☐ No	☐ Yes ☐ No	☐ Yes ☐ No	☐ Yes ☐ No
4. prepare ahead for each class?	☐ Yes ☐ No	☐ Yes ☐ No	☐ Yes ☐ No	☐ Yes ☐ No	☐ Yes ☐ No
5. read previous class notes before each class?	☐ Yes ☐ No	☐ Yes ☐ No	☐ Yes ☐ No	☐ Yes ☐ No	☐ Yes ☐ No
6. pick out key points in lectures?	☐ Yes ☐ No	☐ Yes ☐ No	☐ Yes ☐ No	☐ Yes ☐ No	☐ Yes ☐ No
7. minimize distractions and concentrate on the lecture?	☐ Yes ☐ No	☐ Yes ☐ No	☐ Yes ☐ No	☐ Yes ☐ No	☐ Yes ☐ No
8. feel comfortable using office hours for questions?	☐ Yes ☐ No	☐ Yes ☐ No	☐ Yes ☐ No	☐ Yes ☐ No	☐ Yes ☐ No

9. feel comfortable saying that you don't understand a point?

☐ Yes ☐ Yes ☐ Yes ☐ Yes ☐ Yes
☐ No ☐ No ☐ No ☐ No ☐ No

10. pick out the component parts of a lecture?

☐ Yes ☐ Yes ☐ Yes ☐ Yes ☐ Yes
☐ No ☐ No ☐ No ☐ No ☐ No

11. prevent yourself from tuning-out?

☐ Yes ☐ Yes ☐ Yes ☐ Yes ☐ Yes
☐ No ☐ No ☐ No ☐ No ☐ No

12. use your notes to indicate where you were lost?

☐ Yes ☐ Yes ☐ Yes ☐ Yes ☐ Yes
☐ No ☐ No ☐ No ☐ No ☐ No

Attendance

Regular attendance of lectures is critically important. If you miss a lecture, you may miss many important ingredients that accompany a full understanding of what you are expected to know, or of what you are expected to do. Worse, you may miss assignments, information that is not in the text, or examples that may make your assignments easier. These ingredients are difficult to recapture from a classmate.

Class participation

Every student has at least one question to ask. It may be as simple as "I didn't quite understand. Could you repeat that explanation?" or it may be a question about an unfamiliar symbol or formula. Instructors of calculus are intelligent enough to know that there is no such thing as a *dumb* question. You may think that your question is *dumb*, but others may see it as important. If your class is large, you may feel uncomfortable asking questions during class. This is natural; so you may wish to save your questions for an office visit with your instructor. The first question is always the hardest. So ask your first question of the term early in the term. Visit your instuctor during office hours. Once you do, you will see how easy it is to continue to ask questions during office hours.

How to listen

Picture the face of a friend or an acquaintance. You are recalling only a catalogue of facial features and reconstructing the face from those special features. Usually, the shape of the nose, the color, texture and style of the hair, etc., are all that you need to reconstruct the face that is recognizable as your friend's. It is astonishing: The human brain can construct the whole from very few of the parts! This phenomenon suggests a strategy for listening in the classroom.

Even the best classroom lecture can be effectively reconstructed from a few of its critical parts. Typically, a fifty minute calculus lecture will contain less than a half-dozen, ideas and points that are critical to further understanding of the subject. These ideas and points will be repeated several times during a lecture, once to introduce them and again to clarify them.

Most lectures have at least three clearly defined component parts-procedures, explanations and examples. A procedure provides a way of handling certain types of problems. An explanation is a reason for why certain things are possible and an example is a specific case to focus on.

A good lecturer will prompt you at the transitions between procedure, explanation and example in several possible ways. He or she may noticeably alter voice pitch, resume an emphatic hand gesture, change the size or style of blackboard writing, or repeat the idea while looking intently at the audience. Some lecturers use more clearly defined phrase prompts:

"...and now we move on to..."
"...any questions about this material?"
"...next, I shall explain about..."

With practice you will acquire the skill of recognizing the component parts and use this skill to roughly reconstruct the lecture. Explanations are usually understood only after many encounters with the examples; so give your full attention to the examples.

A lecture on *completing the square* may begin with a polynomial equation of the form $x^2 + bx = c$. Can you identify the transitions between procedure, explanation and example in the following paragraphs?

The procedure begins

Take half of the coefficient of x, square it and add the result to both sides

$$x^2 + bx + \left(\frac{b}{2}\right)^2 = c + \left(\frac{b}{2}\right)^2.$$

The left side then becomes a perfect square and the equation may be rewritten as

$$\left(x + \frac{b}{2}\right)^2 = c + \left(\frac{b}{2}\right)^2.$$

Take the square root of both sides to get

$$x + \frac{b}{2} = \pm\sqrt{c + \frac{b^2}{4}},$$

or

$$x = \frac{-b}{2} \pm \sqrt{c + \frac{b^2}{4}}.$$

An explanation starts here

Our aim was to solve for x in the equation $x^2 + bx = c$ If we could add something to both sides of the equation to make the left side a perfect square of the form

$$(x + \text{number})^2 = \text{another number},$$

then we could take the square root of each side and find x by simplifying the resulting equation.

This is where the example happens

Take the equation $x^2 + 6x = 1$. You may add $\left(\frac{6}{3}\right)^2$ or 9 to both sides to get $x^2 + 6x + 9 = 1 + 9$. This may be rewritten as $(x + 3)^2 = 10$. Take the square root of both sides to get

$$x + 3 = \pm\sqrt{10},$$

or

$$x = -3 \pm \sqrt{10}.$$

What to do when you get lost

Sooner or later you will find a moment when you are lost. The lecture has moved too quickly. You feel that you are hearing the words but not understanding their meaning. What do you do now? Many students encounter this point early in the lecture and *tune out* for the remainder of the lecture. Fortunately, there is enough repetition in a lecture so that you can pick up the points and ideas the second time around.

There are a few strategies to deal with getting lost during a lecture.

1. Try to pick up the thread of the lecture. Do not tune out!

2. If you are brave enough, you may raise your hand and confess to the lecturer that you missed the thread of his or her last remarks. Then politely ask for a short rewinding. You may be surprised; most lecturers are impressed by students who are willing to admit that they do not understand a point. If you do not understand a point, then chances are that your classmates did not understand. Remember that it is normal for you to feel that your classmates understand everything, even though they don't. Of course you should balance the privilege of saying that you do not understand with the risk of appearing obnoxious; do not over-use the privilege!

Don't do this too often. It may become annoying to the class and instructor.

3. Use your notebook to indicate that you are lost at certain points of the lecture. After the lecture, ask the lecturer for material that may help you to understand those points marked in your notebook.

4. See your instructor during his or her next office hour.

Recalling the previous lecture

Consecutive lectures will often overlap. Typically, the lecturer will begin a lecture by recalling the last few minutes of the previous lecture. This is not always done and some lecturers never do it; but, you will find yourself at a big advantage if you have given some thought to the end of the previous lecture.

In recalling the previous lecture, you should pick out critical areas, those ideas that gave you trouble and those areas that forced you to think. Make some notation that will indicate the points to be raised or questions to be asked at the beginning of the next lecture. Visit your instructor during office hours to ask questions that need detailed answers. Instructors are impressed with students who contemplate earlier lectures and who show that they remember what was previously said.

A review of the previous lecture will put you in good shape for understanding the next one.

If you come to a lecture cold, not having reviewed the previous lecture, then you run the risk of loosing the thread at the very beginning of the lecture.

Coordinate your reading with your listening

Understand that the course is made from lectures, readings and exercises. One component without the other would make understanding impossible. You take in information from reading your textbook and listening to lectures; you process this information and merge it with other knowledge by working through many exercises. Attentive listening and clear reading should be coordinated so as to make the merging process easier.

Your notes are the links between your class and your text.

In general, your lecture notes will supplement your text explanations. This means that you will acquire a deeper understanding of the text material by reviewing your notes while you are reading your text; and vice versa, you will have a deeper understanding of your notes and the lecture by reviewing the material in your text.

You have several sources from which to learn the subject-the lectures, the text, your notes, and your thoughts. You should be using all of these sources and coordinating the information that each gives. The lecture notes offer an outline from which to research the topics in the text.

There is a tendency for students to read the text, forgetting that they have lecture notes to guide them; likewise, many students study their lecture notes without paying simultaneous attention to the text.

Tips

- Attend class regularly
- Ask questions early in the term
- Listen for critical points
- If you get lost, indicate it in your notebook and see your instructor after class or during the next office hour.
- Review the previous lecture before going to class
- Coordinate your reading with the lectures

Taking Notes

How good are your notes?

	Now ↓	week 1 ↓	week 2 ↓	week 3 ↓	week 4 ↓
Do you ...					
1. prepare for the lecture by reviewing previous notes?	☐ <u>Yes</u> ☐ <u>No</u>	☐ <u>Yes</u> ☐ <u>No</u>	☐ <u>Yes</u> ☐ <u>No</u>	☐ <u>Yes</u> ☐ <u>No</u>	☐ <u>Yes</u> ☐ <u>No</u>
2. sit where you can see and hear clearly?	☐ <u>Yes</u> ☐ <u>No</u>	☐ <u>Yes</u> ☐ <u>No</u>	☐ <u>Yes</u> ☐ <u>No</u>	☐ <u>Yes</u> ☐ <u>No</u>	☐ <u>Yes</u> ☐ <u>No</u>
3. have a pen or pencil that writes well and feels comfortable?	☐ <u>Yes</u> ☐ <u>No</u>	☐ <u>Yes</u> ☐ <u>No</u>	☐ <u>Yes</u> ☐ <u>No</u>	☐ <u>Yes</u> ☐ <u>No</u>	☐ <u>Yes</u> ☐ <u>No</u>
4. abbreviate words that you write in your notes?	☐ <u>Yes</u> ☐ <u>No</u>	☐ <u>Yes</u> ☐ <u>No</u>	☐ <u>Yes</u> ☐ <u>No</u>	☐ <u>Yes</u> ☐ <u>No</u>	☐ <u>Yes</u> ☐ <u>No</u>
5. write phrases instead of whole sentences?	☐ <u>Yes</u> ☐ <u>No</u>	☐ <u>Yes</u> ☐ <u>No</u>	☐ <u>Yes</u> ☐ <u>No</u>	☐ <u>Yes</u> ☐ <u>No</u>	☐ <u>Yes</u> ☐ <u>No</u>
6. occasionally write words instead of phrases?	☐ <u>Yes</u> ☐ <u>No</u>	☐ <u>Yes</u> ☐ <u>No</u>	☐ <u>Yes</u> ☐ <u>No</u>	☐ <u>Yes</u> ☐ <u>No</u>	☐ <u>Yes</u> ☐ <u>No</u>
7. use symbols for important points and points that you do not understand to ask questions during the lecture or during office hours?	☐ <u>Yes</u> ☐ <u>No</u>	☐ <u>Yes</u> ☐ <u>No</u>	☐ <u>Yes</u> ☐ <u>No</u>	☐ <u>Yes</u> ☐ <u>No</u>	☐ <u>Yes</u> ☐ <u>No</u>

8. write whatever the lecturer writes unless you are told not to?   ☐ Yes ☐ No   ☐ Yes ☐ No   ☐ Yes ☐ No   ☐ Yes ☐ No   ☐ Yes ☐ No

9. ignore irrelevant talk like anecdotes?   ☐ Yes ☐ No   ☐ Yes ☐ No   ☐ Yes ☐ No   ☐ Yes ☐ No   ☐ Yes ☐ No

10. concentrate on the lecture, not on the note-taking?   ☐ Yes ☐ No   ☐ Yes ☐ No   ☐ Yes ☐ No   ☐ Yes ☐ No   ☐ Yes ☐ No

11. copy your notes so they are clear and useful the next day, week or month?   ☐ Yes ☐ No   ☐ Yes ☐ No   ☐ Yes ☐ No   ☐ Yes ☐ No   ☐ Yes ☐ No

12. review your notes just before the next class?   ☐ Yes ☐ No   ☐ Yes ☐ No   ☐ Yes ☐ No   ☐ Yes ☐ No   ☐ Yes ☐ No

What are the important points?

Taking notes is an art. However, like all arts, there are some skills that must be learned.

The most important thing to realize while taking notes at a lecture is that you should never worry about complete sentences. If the lecturer says,

"The quadratic expression $x^2 + (a+b)x + ab$ has factors $(x + a)$ and $(x + b)$," then it is wise to write

"$x^2 + (a+b)x + ab$ factors $(x + a), (x + b)$."

It seems like there is little difference between the two; however, the extra words *The quadratic expression*, and *has factors*, will take at least six seconds of time and may distract you enough to miss the lecturer's next comments. In a fifty minute lecture, you will save significant writing time by making abbreviated notes such as the one

Never write at the expense of listening!

indicated above. Practice making up your own shorthand. You may wish to make symbols to indicate trouble spots or things to remember. The symbol

may suggest a difficult point-it is the French road sign symbol that indicates *dangerous curves ahead*. Here are four other suggested symbols:

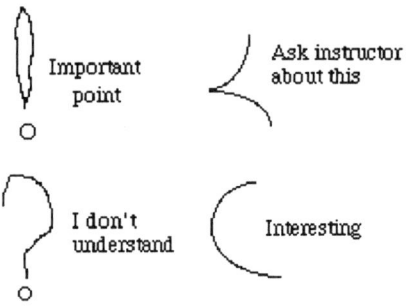

The second most important thing to realize while taking notes is that you should be conservative in your note taking. (If you feel that you must record all of the lecture, then do so with a silent tape recorder-there are some inexpensive, quiet tape recorders on the market that are designed for recording lectures.) So, what do you write and what do you not write? A general rule of thumb is write anything that the lecturer writes (if a blackboard or overhead projector is used). Write down the examples. Even if you do not have time to write anything said about them, you will have something to work on when you review the lecture. In addition to this, you should try to capture any comments that you think express the lecturer's original points of view. Trust your instinct on this, it usually works well.

How much to write

Do not write at the expense of listening.

Naturally, every lecture is different; so, the number of pages that you use will vary from lecture to lecture. Moreover, what you write, how large you write and what you will need to write will depend on you. Some people will recall material from other classes and will not have the need to review as much as others.

Relax while taking notes and understand that you may adjust your personal style of taking notes at the next lecture. If you have not included enough

16

detail for one lecture, try to increase the amount of detail at the next.

How to use what you have written

Notes are meant to be read. Many students fall into the habit of taking notes for the sake of taking notes. They artfully accumulate pages of notes without ever looking at them. It is important to revise a fresh set of notes from the rough sketches that you made at the time of the lecture. This should be done as soon as possible, for otherwise you are likely to forget the meanings of the sketches.

Once the notes are revised, they should be read alongside the text. If a page of notes is about continuous functions then it should be read in conjunction with the sections of the text that discuss continuous functions. The lecturer discusses things in more detail than your text can. Moreover, your notes represent material that the lecturer considers important. This means that he or she expects you to put more emphasis into studying the topics referred to in your notes.

A good set of notes will be extremely helpful when it comes time to study for an exam. Your notes will remind you of your weaknesses. They will remind you of the focal points that were important to the lecturer and therefore should be important to you.

Sample notes

Forget about correct grammar when taking notes.

Avoid complete sentences and use abbreviation

The usefulness of your notes will depend on how well you have taken them. A sample from a lecture on continuous functions follows. It shows how abbreviations and symbols can be used to speed the writing and allow more time for concentration during the lecture. The notes in this sample form a collection of phrases, not sentences. Notes are rarely composed of complete sentences; all grammar is excused when you are taking notes. Don't even try to use correct grammar when taking notes.

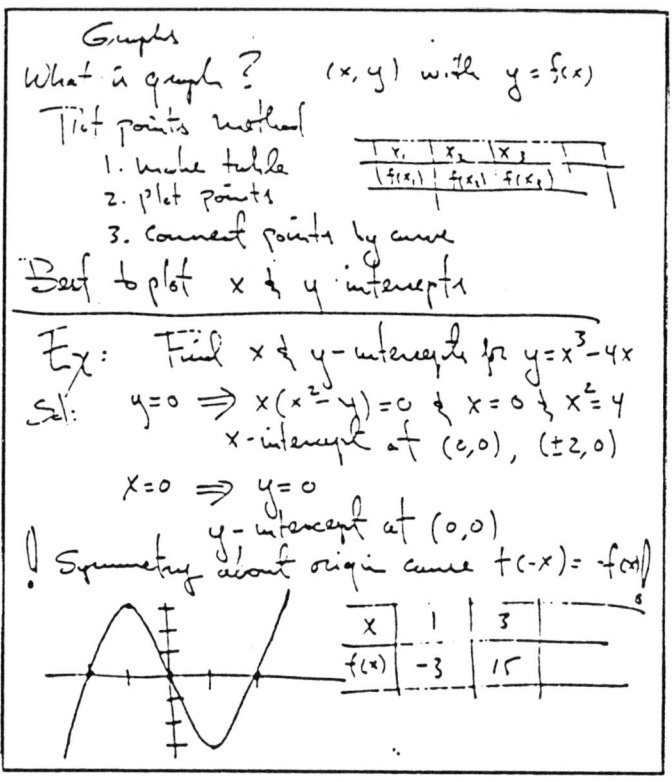

Ex denotes example

Sol denotes solution and

$\Rightarrow$ denotes implies.

Tips

- Never worry about completing a sentence while taking notes
- Abbreviate, using marks, symbols and rough sketches
- Write down examples used in class
- Rewrite notes soon after class to make them clear for when you need them
- Review your notes along with your text, topic for topic
- Use your notes in preparing for classes and quizzes

Reading the Text 4

How carefully do you read?

	Now ↓	week 1 ↓	week 2 ↓	week 3 ↓	week 4 ↓
Do you ...					
1. read the text before trying to work the exercises?	☐ Yes ☐ No	☐ Yes ☐ No	☐ Yes ☐ No	☐ Yes ☐ No	☐ Yes ☐ No
2. try to understand the material, rather than force it into short term memory?	☐ Yes ☐ No	☐ Yes ☐ No	☐ Yes ☐ No	☐ Yes ☐ No	☐ Yes ☐ No
3. try to understand boxed formulas, rather than memorize them?	☐ Yes ☐ No	☐ Yes ☐ No	☐ Yes ☐ No	☐ Yes ☐ No	☐ Yes ☐ No
4. focus your thought on understanding relationships between equations?	☐ Yes ☐ No	☐ Yes ☐ No	☐ Yes ☐ No	☐ Yes ☐ No	☐ Yes ☐ No
5. make sure that you are not distracted?	☐ Yes ☐ No	☐ Yes ☐ No	☐ Yes ☐ No	☐ Yes ☐ No	☐ Yes ☐ No
6. read each section twice, once fast and once again for detail?	☐ Yes ☐ No	☐ Yes ☐ No	☐ Yes ☐ No	☐ Yes ☐ No	☐ Yes ☐ No
7. look up unfamiliar terms?	☐ Yes ☐ No	☐ Yes ☐ No	☐ Yes ☐ No	☐ Yes ☐ No	☐ Yes ☐ No
8. use the margins of the text for notes to yourself?	☐ Yes ☐ No	☐ Yes ☐ No	☐ Yes ☐ No	☐ Yes ☐ No	☐ Yes ☐ No

9. highlight or underline only when absolutely necessary?

| ☐ Yes | ☐ Yes | ☐ Yes | ☐ Yes | ☐ Yes |
| ☐ No | ☐ No | ☐ No | ☐ No | ☐ No |

10. pay particular attention to examples and try to work them out for yourself?

| ☐ Yes | ☐ Yes | ☐ Yes | ☐ Yes | ☐ Yes |
| ☐ No | ☐ No | ☐ No | ☐ No | ☐ No |

11. mark parts that you should review or return to for more understanding?

| ☐ Yes | ☐ Yes | ☐ Yes | ☐ Yes | ☐ Yes |
| ☐ No | ☐ No | ☐ No | ☐ No | ☐ No |

12. coordinate your text material with your class notes and then read ahead?

| ☐ Yes | ☐ Yes | ☐ Yes | ☐ Yes | ☐ Yes |
| ☐ No | ☐ No | ☐ No | ☐ No | ☐ No |

How to begin

Try to get a broad understanding of what you are doing. Read the section before trying the exercises.

Many students mistakenly believe that they are saving study time by beginning the exercises before reading the accompanying section of the text. They go directly to the exercises, and look up how to do them afterwards. The problem is that you are forcing yourself to rely on short-term memory, rather than on a broader understanding of the material as a whole. Short-term memory works for the short-term. The biggest problem with relying on short-term memory is that it is a very inefficient way of studying a subject that takes more than a year to understand. Each time you force a *boxed* formula or *boxed* definition into short-term memory, solely to use it for an immediate exercise, you will find it necessary to look up that formula or definition again and again for future exercises.

Read the section twice – the first time quickly, the second slowly.

It is good practice to read a section twice, once for a general feeling of the content and a second time for the details. The first reading should be fast. You want to get an overall *feeling* for what is being presented-no details. You will not understand much from this fast reading, the objective is to see the symbols, theorems or examples once before learning what they mean and how they fit together.

The second reading should be slow and careful. You should have a pencil in hand to rewrite important formulas, theorems or diagrams. Rewriting forces

Write things down as you read. Writing provides mysterious links between reading and thinking.

information into memory, if not altogether into understanding. Write as much as you can, for the writing process has a mysterious connection with mental processing and absorption.

Look up any unfamiliar terms. You may have to do this repeatedly for each term before the meaning sticks. It is time-consuming but worth the effort. Use the index at the back of your textbook to find the pages on which the terms appear.

Using the margins

This booklet has wide margins throughout. Wide margins provide two advantages. One is the ease with which material can be scanned and read; the other is that notes and comments can be clearly added by the reader. Almost every calculus text has a wide margin. These margins are made to be marked up. If you are worried about the resale value of your text, don't. Think about the cost of the course itself. At a typical college, that cost is between $300 and $500, certainly more than ten times what you would get for the used book at the end of the year. By comparison with the price of a calculus course, the text is very inexpensive.

You should be marking the margins of this book as well. Then, you could use the marginal marks and comments to briefly review what you have learned.

A well marked margin is an excellent tool for studying for exams. You should mark those parts of the text that you find especially useful to review when the time comes for review. Mark the places where you are having trouble and the places where there is material that you find useful. In studying for your exams you will find that there is no need to reread the entire sections; rather, follow your margin notes and let them suggest which areas need to be emphasized in your review.

Highlighting and how to use it

There is a limit to the amount of highlighting that should be used. Highlighting is a good tool for marking important formulas or passages that you may wish to refer to when you are ready to review your studies. However, if you over highlight, you are not limiting the amount of material reserved for reviewing, and therefore are missing the whole point of highlighting.

Organizing your reading

For many students, it is best to work out an example in the text before the detailed (second) reading of the section. Try to work the example without help from the text. Use the text as a guide. In this way you may become more motivated to read the section.

Read sections of the text in the way you feel most comfortable; however, be sure to organize your reading. It is not always necessary to read each section from beginning to end and in that order. You may find an example that is interesting and follow the example backwards until you understand it. Remember that the reason for reading the section is to understand it, not simply to read it in record time. Motivation is the most important ingredient in studying. If you begin with an example that you understand, then you are more likely to be motivated to continue your reading. The textbook itself cannot jump around as you can. It must conform to a certain prescribed format.

If you do not understand a sentence, try reading the next sentence; often, subsequent sentences will explain earlier ones. If, after reading several sentences, you still do not understand what you are reading, then move to an earlier paragraph; perhaps you skipped over something critically important, something that is needed for understanding later paragraphs. If you feel stuck after spending-what you feel to be-too much time on the paragraph, then skip to the next example or the next topic within the section.

Always remember that a worked example and a little thought can go a long way in bringing out a clearer understanding of the topic.

Tips

- Roughly read a section for feeling rather than for details
- Read the section once again. This time slowly and carefully
- Write down important formulas, theorems or diagrams
- Look up unfamiliar terms
- Use the margins of the text to indicate trouble spots and places for review
- Highlight sparingly
- Try to work out some examples with help from the text

Working through a Problem 5

Are you prepared to solve problems?

	Now ↓	week 1 ↓	week 2 ↓	week 3 ↓	week 4 ↓
Do you ...					
1. carefully read every part of each exercise?	☐ <u>Yes</u> ☐ <u>No</u>	☐ <u>Yes</u> ☐ <u>No</u>	☐ <u>Yes</u> ☐ <u>No</u>	☐ <u>Yes</u> ☐ <u>No</u>	☐ <u>Yes</u> ☐ <u>No</u>
2. understand what each word and equation means?	☐ <u>Yes</u> ☐ <u>No</u>	☐ <u>Yes</u> ☐ <u>No</u>	☐ <u>Yes</u> ☐ <u>No</u>	☐ <u>Yes</u> ☐ <u>No</u>	☐ <u>Yes</u> ☐ <u>No</u>
3. isolate questions?	☐ <u>Yes</u> ☐ <u>No</u>	☐ <u>Yes</u> ☐ <u>No</u>	☐ <u>Yes</u> ☐ <u>No</u>	☐ <u>Yes</u> ☐ <u>No</u>	☐ <u>Yes</u> ☐ <u>No</u>
4. try to simplify questions?	☐ <u>Yes</u> ☐ <u>No</u>	☐ <u>Yes</u> ☐ <u>No</u>	☐ <u>Yes</u> ☐ <u>No</u>	☐ <u>Yes</u> ☐ <u>No</u>	☐ <u>Yes</u> ☐ <u>No</u>
5. break questions down into simpler, component questions?	☐ <u>Yes</u> ☐ <u>No</u>	☐ <u>Yes</u> ☐ <u>No</u>	☐ <u>Yes</u> ☐ <u>No</u>	☐ <u>Yes</u> ☐ <u>No</u>	☐ <u>Yes</u> ☐ <u>No</u>
6. keep organized worksheets?	☐ <u>Yes</u> ☐ <u>No</u>	☐ <u>Yes</u> ☐ <u>No</u>	☐ <u>Yes</u> ☐ <u>No</u>	☐ <u>Yes</u> ☐ <u>No</u>	☐ <u>Yes</u> ☐ <u>No</u>
7. write down what is known and what is not known?	☐ <u>Yes</u> ☐ <u>No</u>	☐ <u>Yes</u> ☐ <u>No</u>	☐ <u>Yes</u> ☐ <u>No</u>	☐ <u>Yes</u> ☐ <u>No</u>	☐ <u>Yes</u> ☐ <u>No</u>
8. try to interpret the meanings of your answers?	☐ <u>Yes</u> ☐ <u>No</u>	☐ <u>Yes</u> ☐ <u>No</u>	☐ <u>Yes</u> ☐ <u>No</u>	☐ <u>Yes</u> ☐ <u>No</u>	☐ <u>Yes</u> ☐ <u>No</u>

9. test your answers to see if they make sense? ☐ <u>Yes</u> ☐ <u>Yes</u> ☐ <u>Yes</u> ☐ <u>Yes</u> ☐ <u>Yes</u>
☐ <u>No</u> ☐ <u>No</u> ☐ <u>No</u> ☐ <u>No</u> ☐ <u>No</u>

10. discuss your work with classmates? ☐ <u>Yes</u> ☐ <u>Yes</u> ☐ <u>Yes</u> ☐ <u>Yes</u> ☐ <u>Yes</u>
☐ <u>No</u> ☐ <u>No</u> ☐ <u>No</u> ☐ <u>No</u> ☐ <u>No</u>

11. speak to your instructor about material that you do not understand? ☐ <u>Yes</u> ☐ <u>Yes</u> ☐ <u>Yes</u> ☐ <u>Yes</u> ☐ <u>Yes</u>
☐ <u>No</u> ☐ <u>No</u> ☐ <u>No</u> ☐ <u>No</u> ☐ <u>No</u>

12. keep a record of unsolved problems and return to them at a later time? ☐ <u>Yes</u> ☐ <u>Yes</u> ☐ <u>Yes</u> ☐ <u>Yes</u> ☐ <u>Yes</u>
☐ <u>No</u> ☐ <u>No</u> ☐ <u>No</u> ☐ <u>No</u> ☐ <u>No</u>

Organizing your thoughts

The trick to good problem solving is in organization of thought. Many problems fall into categories or types which can be ordered by complexity or by dependence. For example: to solve the equation

$$x^2 + 2x + 1 = 0,$$

it would be helpful to know how to solve

$$(x + 1)^2 = 0.$$

This is the factored form of the first equation.

The second equation is easier to solve than the first; and the solutions are found by taking the square roots of both sides of the last equation.

Notice that we solved the original equation by simplifying the question-the second equation is the factored form of the first. Many exercises are solved in exactly this way. You first look for a simplification of the statement of the exercise, then work on the simplification. When attacking a problem, always ask yourself if there is an easier way of stating the exercise.

Isolating the question

You may be eager to get through your assignments in a limited amount of time. In doing so, you may neglect to read the problems carefully and give thought to the problems themselves. Jumping to the solution without thinking about the problem may lead to an unnecessarily complicated solution. Complicated solutions take more time, are more vulnerable to error and are not easy to check. On an exam, you should look for the simplest solution so that it can be checked and written down in the shortest amount of time. For example, on a quiz that asks for the minimum value of y in the equation

$$y = (3x + 3)^2,$$

You may want to notice that this is a perfect square which can never have a value less than zero.

you may not recognize that the right side of the above equation is a square which cannot be negative. Since y = 0 when x = -1, the minimum value of y must be 0. You may arrive at the same result by using some unnecessarily complicated procedures of calculus, thereby loosing considerable time and not completing the exam.

Moral: Think about the problem before writing your first thought about a solution.

Developing strategies

A strategy is a plan for achieving an end. To have a strategy requires an understanding of the end. Before beginning the solution to a problem it is important to have the end of the solution in mind. For example, in the problem of finding the minimum value of y in the equation

$$y = (3x + 3)^2,$$

This is a perfect square. A perfect square cannot take on a value less than zero. Can it actually be zero?

the end result is that the value of y that you are looking for must be the smallest value of y that the equation can support. You see, the focus of attention should be on the fact that, whatever the answer is, it must be the same as $(3x + 3)^2$ for some x and also the smallest that it can possibly be.

Simplifying the problem

The most effective strategy in problem solving is simplification. Take our previous task of finding the smallest value of y for the equation

$$y = (3x + 3)^2.$$

If you do not know how to proceed, then present yourself with a simpler problem. You may ask yourself to find the smallest value of y for the equation

Rather than attacking the more complicated problem, we replace it with a simpler one.

$$y = x^2,$$

notice that y is smallest when x is zero, then return to the first problem with the idea that y is smallest when 3x + 3 is zero (or when x = -1).

Every problem you encounter has a simplification. If you develop a habit of looking for simplifications, problem-solving routines will become almost automatic.

Breaking down the problem

Another way to simplify a problem is to break it down into component parts, each part being a simpler problem. This technique is developed through practice.

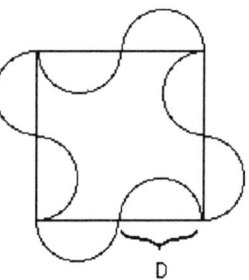

D

There are eight semicircles, each having a circumference $\pi \frac{D}{2}$

For example: Suppose that you are asked to find the area of the square in the figure on the previous page, when you know that the length of the curved figure is 16π. To solve this problem, first solve the *sub*problem of finding the length of the diameter D of one of the semicircles. To do this, count the number of semicircles (8), and note that each has a circumference equal to $\pi \frac{D}{2}$. [because the circumference of a complete circle is 2πr, where r is the radius and therefore the circumference of a semicircle is πD.]

This tells you that $8\left(\pi \frac{D}{2}\right) = 16\pi$, or that D = 4. This solves the *sub*problem. Returning to the original problem you notice that each side has the length of two diameters (2D), or 8 units. The area of the square is therefore 64 units.

All mathematics is structured by a dependency ordering-calculus depends on algebra, while algebra, in turn, depends on arithmetic. Individual problems generally have a similar dependency structure-a problem at one level of calculus will depend on a problem at a lower level of calculus. Finding the simpler subproblems is a key to solving the higher level problems.

Organizing your worksheet

A disorganized worksheet will direct your thoughts along misguided paths and make your solutions vulnerable to possible errors. Write legibly and clearly; and-no less important-place your writing in strategic positions on your paper. A long string of calculations without commentary will be difficult to read and study; moreover, your understanding of your own solution may become muddled.

Consider the following example: A train traveling 60 miles per hour for 20 minutes covers the same distance as a train traveling 40 miles per hour for how many hours? To solve this problem we must compute the distance that the first train travels in 20 minutes. Then compute how long it will take the second train to cover that same distance. Take a look at the two worksheets on the next page. If your written solution looks like the sheet on the left it is correct, but muddled. Not only will you not have a clear picture of the solution in your mind, but too, your notes will be useless for reviewing at a later date. The solution on the right is clearer.

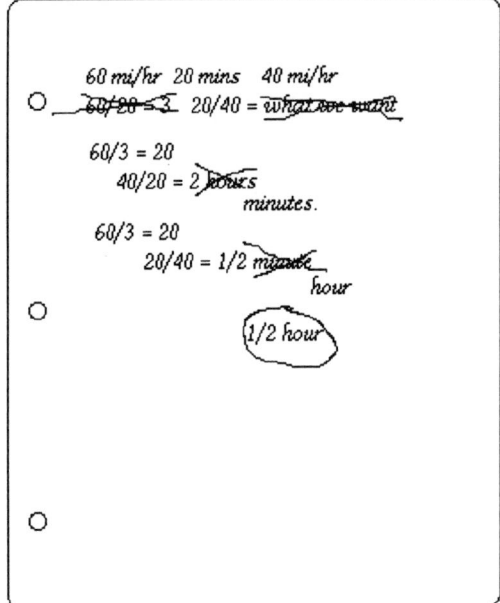

Interpreting your results

Once a problem has been solved, it is important to reflect upon the meaning of its solution.

- Is it reasonable?

- Is it possible?

- Does it make sense?

For example, is the solution of our train problem reasonable? Yes, it should take longer for the second train to travel the same distance as the first, because the second train is slower than the first. But how much longer? The second train is traveling at two-thirds the speed of the first; so is it reasonable that it takes it three-halves the amount of time to travel the same distance? If so, how long will it take the second train to go the distance that the first train covers in one hour? Shouldn't it be three-halves of an hour? A quick computation tells us that it does.

This is a very important point!

The extra questions that surround the solution add significantly to the understanding of the problem.. A little investigation of the problem revealed the true spirit of the question. It is a question about proportions-the second train is 2/3 the speed of the first and therefore takes 3/2 the time to achieve the same distance, no matter what the distance is.

Develop good habits

The skills discussed in this section are not easy to implement at all times. If you are aware that thoughts can be organized, problems broken down and simplified, and that solutions should be interpreted, then you should try these techniques from time to time. Habits develop slowly, they need nudges from time to time.

Work with classmates

Mathematics is meant to be communicated. Not only is it more fun to work on mathematics together with someone, but it is also more rewarding. By communicating what you do know and by asking about what you don't know, you will learn quickly and efficiently.

Use your instructor

Remember that your instructor is there to help you. It is always best to solve problems on your own; but if you do get stuck do not hesitate to ask your instructor for help.

Tips

- Think about the problem before jumping to a solution
- Isolate the question
- Develop a strategy such as simplifying the problem or breaking it into smaller problems
- Organize your work so that you don't get confused
- Check and interpret your results
- Work with your classmates. Share what you know and don't know
- Ask for help when you are stuck

Preparation of Assignments

Do you work on assignments effectively?

	Now ↓	week 1 ↓	week 2 ↓	week 3 ↓	week 4 ↓
Do you ...					
1. keep assignments and directions in a separate section of your notebook	☐ Yes ☐ No	☐ Yes ☐ No	☐ Yes ☐ No	☐ Yes ☐ No	☐ Yes ☐ No
2. understand your assignment before leaving the classroom?	☐ Yes ☐ No	☐ Yes ☐ No	☐ Yes ☐ No	☐ Yes ☐ No	☐ Yes ☐ No
3. work with neat paper and sharp pencil or good pen?	☐ Yes ☐ No	☐ Yes ☐ No	☐ Yes ☐ No	☐ Yes ☐ No	☐ Yes ☐ No
4. indent and organize your writing on the page?	☐ Yes ☐ No	☐ Yes ☐ No	☐ Yes ☐ No	☐ Yes ☐ No	☐ Yes ☐ No
5. sketch useful diagrams and graphs clearly and neatly?	☐ Yes ☐ No	☐ Yes ☐ No	☐ Yes ☐ No	☐ Yes ☐ No	☐ Yes ☐ No
6. work calculations out on a scratch sheet?	☐ Yes ☐ No	☐ Yes ☐ No	☐ Yes ☐ No	☐ Yes ☐ No	☐ Yes ☐ No
7. carefully check your work and answers?	☐ Yes ☐ No	☐ Yes ☐ No	☐ Yes ☐ No	☐ Yes ☐ No	☐ Yes ☐ No
8. rewrite pages that do not look neat and organized?	☐ Yes ☐ No	☐ Yes ☐ No	☐ Yes ☐ No	☐ Yes ☐ No	☐ Yes ☐ No

9. indicate when you could not do a problem or get the correct answer?

☐ <u>Yes</u> ☐ <u>Yes</u> ☐ <u>Yes</u> ☐ <u>Yes</u> ☐ <u>Yes</u>
☐ <u>No</u> ☐ <u>No</u> ☐ <u>No</u> ☐ <u>No</u> ☐ <u>No</u>

10. not bluff your way to an answer, even though it may be correct?

☐ <u>Yes</u> ☐ <u>Yes</u> ☐ <u>Yes</u> ☐ <u>Yes</u> ☐ <u>Yes</u>
☐ <u>No</u> ☐ <u>No</u> ☐ <u>No</u> ☐ <u>No</u> ☐ <u>No</u>

11. carefully check your work and answers?

☐ <u>Yes</u> ☐ <u>Yes</u> ☐ <u>Yes</u> ☐ <u>Yes</u> ☐ <u>Yes</u>
☐ <u>No</u> ☐ <u>No</u> ☐ <u>No</u> ☐ <u>No</u> ☐ <u>No</u>

12. remember to hand your assignments in when due?

☐ <u>Yes</u> ☐ <u>Yes</u> ☐ <u>Yes</u> ☐ <u>Yes</u> ☐ <u>Yes</u>
☐ <u>No</u> ☐ <u>No</u> ☐ <u>No</u> ☐ <u>No</u> ☐ <u>No</u>

Clarification of what is expected

You should have a clear understanding of your next assignment before you leave your classroom. Your instructor or a reliable classmate is your best source for what to do next. Once you leave the classroom, it may take valuable time to find a knowledgeable person to help you understand your assignment.

Presentation of your work

A neat, well presented assignment is a head-start for a better grade. A grader interprets neatness and organization as polite respect for his or her job. Neatly organized papers are easier to read and therefore take less time to grade; all graders appreciate this, even if the solutions are erroneous.

Page layout

Indentation and positioning on the page makes a difference in the way the page is read. Begin with your own system of indentation and page layout. One possible layout is a two column page, the left reserved for important explanations of the solution, the right for computations, annotations and general brainstorming ideas. Be sure to separate independent calculations that are routine and unenlightening.

Drawing useful diagrams and graphs

Many mathematical concepts are difficult to express in normal language. It may be better to learn these concepts through pictures or diagrams.

There are techniques for making accurate, presentable pictures which you should know about. The most important one is what we might call the *point* technique. The *point* technique has three steps:

1. Imagine the completed picture.

2. Pick several strategic points on your imagined picture and draw them.

3. With pencil, lightly connect the points. Use the imagined picture as your model. Erase your lightly drawn outline, if you are not satisfied with it.

4. When you are satisfied with your pencil outline, darken it.

The following sequence of sketches illustrate the technique of drawing a cube.

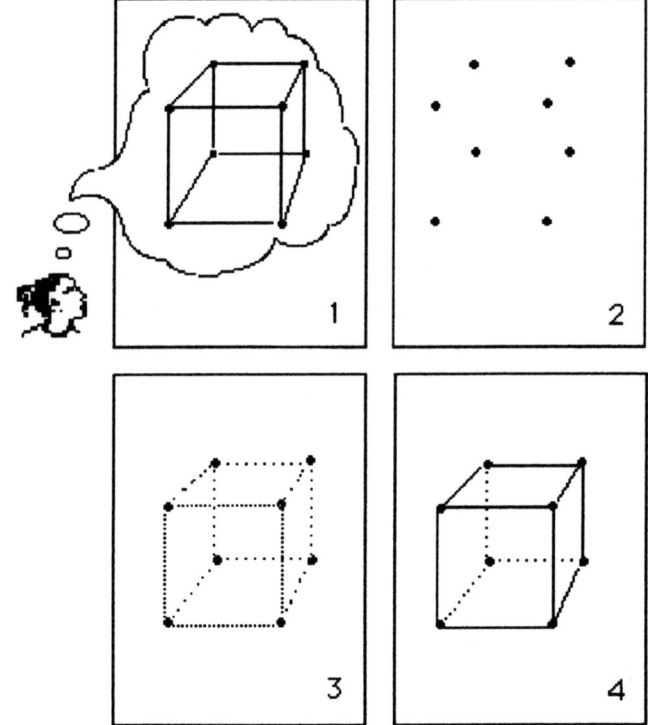

Displaying the answers

Graders look for correct procedures as well as correct answers. If an answer is correct, the grader will scan the procedure to make sure that a proper line of thought lead to the correct answer. If the answer is incorrect, the grader will read through the procedure to point out your errors, usually giving partial credit for your line-of-thought.

In all cases, the answer should be clearly displayed, so the grader can decide on how carefully he or she must read through the solution. Circle the final answer so the grader does not have to search through the pages to decide on which expression is the final result.

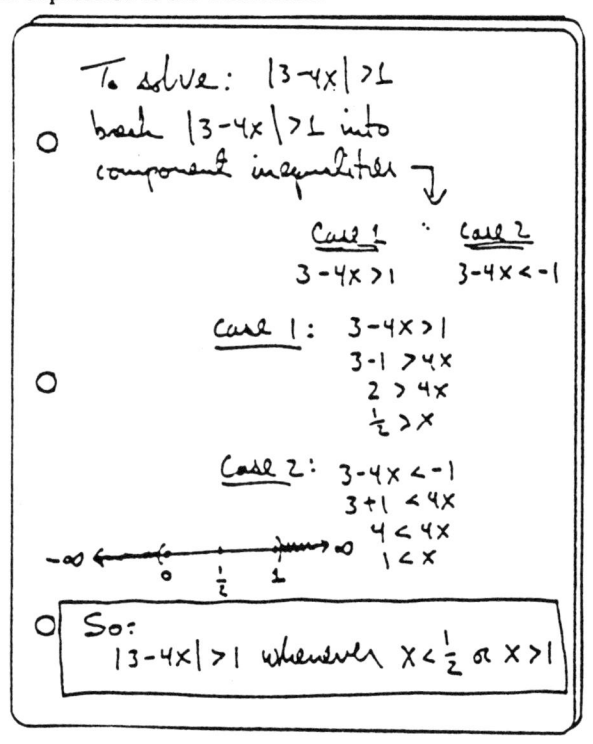

[Handwritten example:]

To solve: $|3-4x| > 1$

○ break $|3-4x| > 1$ into component inequalities ↓

      Case 1   Case 2
      $3-4x > 1$ $3-4x < -1$

Case 1: $3-4x > 1$
$3-1 > 4x$
$2 > 4x$
$\frac{1}{2} > x$

○ Case 2: $3-4x < -1$
$3+1 < 4x$
$4 < 4x$
$1 < x$

[Number line from $-\infty$ to ∞ marked at $\frac{1}{2}$ and 1]

○ So:
$|3-4x| > 1$ whenever $x < \frac{1}{2}$ or $x > 1$

Tips

- Understand your assignment when it is assigned
- Present your work neatly
- Sketch all useful diagrams and graphs
- Clearly display the answers

Preparing for Quizzes and taking Exams 7

Do you know how to prepare for an exam?

	Now	week 1	week 2	week 3	week 4
Do you ...	↓	↓	↓	↓	↓
1. know in advance what may be covered on an exam? _____	☐ <u>Yes</u> ☐ <u>No</u>	☐ <u>Yes</u> ☐ <u>No</u>	☐ <u>Yes</u> ☐ <u>No</u>	☐ <u>Yes</u> ☐ <u>No</u>	☐ <u>Yes</u> ☐ <u>No</u>
2. have a sense of what is likely to be asked on an exam? _____	☐ <u>Yes</u> ☐ <u>No</u>	☐ <u>Yes</u> ☐ <u>No</u>	☐ <u>Yes</u> ☐ <u>No</u>	☐ <u>Yes</u> ☐ <u>No</u>	☐ <u>Yes</u> ☐ <u>No</u>
3. know what you are expected to memorize? _____	☐ <u>Yes</u> ☐ <u>No</u>	☐ <u>Yes</u> ☐ <u>No</u>	☐ <u>Yes</u> ☐ <u>No</u>	☐ <u>Yes</u> ☐ <u>No</u>	☐ <u>Yes</u> ☐ <u>No</u>
4. use your notes when preparing for an exam? _____	☐ <u>Yes</u> ☐ <u>No</u>	☐ <u>Yes</u> ☐ <u>No</u>	☐ <u>Yes</u> ☐ <u>No</u>	☐ <u>Yes</u> ☐ <u>No</u>	☐ <u>Yes</u> ☐ <u>No</u>
5. use your textbook when preparing for an exam? _____	☐ <u>Yes</u> ☐ <u>No</u>	☐ <u>Yes</u> ☐ <u>No</u>	☐ <u>Yes</u> ☐ <u>No</u>	☐ <u>Yes</u> ☐ <u>No</u>	☐ <u>Yes</u> ☐ <u>No</u>
6. use returned assignments when preparing for an exam? _____	☐ <u>Yes</u> ☐ <u>No</u>	☐ <u>Yes</u> ☐ <u>No</u>	☐ <u>Yes</u> ☐ <u>No</u>	☐ <u>Yes</u> ☐ <u>No</u>	☐ <u>Yes</u> ☐ <u>No</u>
7. assess your strengths and weaknesses? _____	☐ <u>Yes</u> ☐ <u>No</u>	☐ <u>Yes</u> ☐ <u>No</u>	☐ <u>Yes</u> ☐ <u>No</u>	☐ <u>Yes</u> ☐ <u>No</u>	☐ <u>Yes</u> ☐ <u>No</u>
8. work on your weaknesses in preparation for an exam? _____	☐ <u>Yes</u> ☐ <u>No</u>	☐ <u>Yes</u> ☐ <u>No</u>	☐ <u>Yes</u> ☐ <u>No</u>	☐ <u>Yes</u> ☐ <u>No</u>	☐ <u>Yes</u> ☐ <u>No</u>

9. get help when you need it?

☐ Yes ☐ Yes ☐ Yes ☐ Yes ☐ Yes
☐ No ☐ No ☐ No ☐ No ☐ No

10. talk with classmates about what they expect to be asked on an exam?

☐ Yes ☐ Yes ☐ Yes ☐ Yes ☐ Yes
☐ No ☐ No ☐ No ☐ No ☐ No

11. space your exam preparation time so that you do not have to cram the night before?

☐ Yes ☐ Yes ☐ Yes ☐ Yes ☐ Yes
☐ No ☐ No ☐ No ☐ No ☐ No

12. get sufficient rest the night before an exam?

☐ Yes ☐ Yes ☐ Yes ☐ Yes ☐ Yes
☐ No ☐ No ☐ No ☐ No ☐ No

Preparation

Preparation should begin with answers to these basic questions:

1. <u>What material will be covered on the quiz or exam?</u>
   The instructor should give a general idea of what will be covered.

2. <u>What is likely to be asked?</u>
   You will have a better answer to this after the second or third quiz; but remember, quiz problems are generally short questions, not questions that involve five sheets of messy calculations.

3. <u>What material should be memorized?</u>
   In general, if you have worked with an equation, an identity, or an expression over a period of time, you will automatically remember it. If you have not extensively used material that is likely to be on the exam, then memorize it. You may ask the instructor about any important formulas that are necessary to memorize.

Using your notes

Your notebook is your guide to studying for quizzes and exams. If you did an adequate job of taking and rewriting your notes, then you should have some indication of what material was covered, what gave you trouble, and what examples to understand. Use the notebook to lead you to other sources of preparation. Formulas, identities and important information could be xeroxed from your text and pasted onto cards or into your notebook. They will provide a quick handy reference for your test review.

Using your textbook

Once you know what material is to be covered on a quiz, review that material in the text, paying particular attention to your own markings and margin comments. Understand as many examples as you can. Work through as many exercises as you have time for. Chapter review exercises offer the best preparation for quizzes, since they are not obviously tied to specific sections of the text. Most calculus texts provide an ample selection of review problems at the end of each chapter.

Using returned assignments

Returned assignments are important components of your quiz preparation material. Here is where your weaknesses will show. All weaknesses must be carefully recognized so that steps can be taken to overcome them.

Don't be discouraged by the grader's marks on returned assignments. Those marks are there to help you discover your weaknesses.

The ultimate purpose of an assignment is to strengthen your skills, while the ultimate purpose of a quiz or exam is to measure your progress.

Students are generally bashful about receiving graded assignments. The first thing they see is a grade, the second is the grader's comments in the margins. These comments often translate into embarrassment or discouragement, when, in fact, the grader simply meant to help the student along the more correct path. Instead of being discouraged by the comments, use them in the way they are intended-as indications of the areas that you need to strengthen. The grades of assignments usually do not count as heavily as those of quizzes or exams.

If your assignments are not collected, ask yourself if you feel comfortable with the material. Do you understand your own solutions? Are you sure that your answers are correct? If the answer is negative, then ask your instructor to look over your work. He or she would be very happy to comment on occasional work that you bring forward. That is an integral part of his or her job!

Using computers and video tutorials

Computer drill and practice programs can be very helpful in preparing you for quizzes, especially if the programs give you hints and refer you to material in the text whenever your answers are incorrect. Ask your instructor about such programs. Video tutorials are useful, particularly if you have trouble following details in a lecture setting, or if you have missed several classes due to illness. They can provide additional examples and perspectives. A video tutorial can be paused, rewound and repeated to accommodate your pace.

Assessing your strengths and weaknesses

Preparing for an exam requires an honest assessment of your weaknesses as well as your strengths. Knowing your strengths gives you confidence, which is the primary ingredient for success on a quiz or exam. Knowing your weaknesses, and everybody has some, gives you a chance to work on them. Quiz preparation becomes easy, once you have a clear idea of where your energies should be concentrated.

Getting help

An instructor is not simply a lecturer. He or she is at your service to help you understand the subject. That usually requires lectures and private explanations. Never hesitate to ask questions in class or out. If you are nervous about the question, then write it down and ask it after class. Just remember that many of your classmates are equally as nervous and probably would like to ask the same question. They will be grateful, if you ask it.

Keep in mind that your instructor was once a student too. He or she will share your concerns.

The first question is always the hardest to ask, in class or out. Once you have asked one question, you will find it easier to ask the next. If you establish your credentials early-as someone who can ask questions-then you will open a door between you, your classmates and your instructor. See your instructor early in the term. Establish a connection by visiting him or her during office hours. Let him or her know of your concerns, difficulties or even satisfaction. Be polite but relax-he or she is a human being concerned with your success in the course. That is part of his or her job. If half the term passes and you have not asked a single question, then it will become psychologically very difficult for you to ask questions.

Friendly advice

Do not stay up late the night before an exam. It may seem like a good idea at the time, but fatigue generally limits concentration and causes more anxiety during the exam. Remember that nervousness during an exam will obstruct creative thinking. So, get plenty of rest the night before an exam. Be sure to come early to the exam so that you will have time to take a few moderately deep relaxing breaths before the exam.

Tips

- Find out what will be covered
- Think about what may be asked
- Review old quizzes and notes
- Check through the margins of your text for markings that indicate trouble spots
- Assess your weaknesses honestly and work on strengthening them

At the Exam 8

Do you know how to take an exam?

Do you ...	Now ↓	week 1 ↓	week 2 ↓	week 3 ↓	week 4 ↓
1. try to relax?	☐ <u>Yes</u> ☐ <u>No</u>	☐ <u>Yes</u> ☐ <u>No</u>	☐ <u>Yes</u> ☐ <u>No</u>	☐ <u>Yes</u> ☐ <u>No</u>	☐ <u>Yes</u> ☐ <u>No</u>
2. read any instructions before beginning?	☐ <u>Yes</u> ☐ <u>No</u>	☐ <u>Yes</u> ☐ <u>No</u>	☐ <u>Yes</u> ☐ <u>No</u>	☐ <u>Yes</u> ☐ <u>No</u>	☐ <u>Yes</u> ☐ <u>No</u>
3. fairly divide the amount of time that you have between problems?	☐ <u>Yes</u> ☐ <u>No</u>	☐ <u>Yes</u> ☐ <u>No</u>	☐ <u>Yes</u> ☐ <u>No</u>	☐ <u>Yes</u> ☐ <u>No</u>	☐ <u>Yes</u> ☐ <u>No</u>
4. scan the exam for familiar questions and do them first?	☐ <u>Yes</u> ☐ <u>No</u>	☐ <u>Yes</u> ☐ <u>No</u>	☐ <u>Yes</u> ☐ <u>No</u>	☐ <u>Yes</u> ☐ <u>No</u>	☐ <u>Yes</u> ☐ <u>No</u>
5. try any questions that are easy first?	☐ <u>Yes</u> ☐ <u>No</u>	☐ <u>Yes</u> ☐ <u>No</u>	☐ <u>Yes</u> ☐ <u>No</u>	☐ <u>Yes</u> ☐ <u>No</u>	☐ <u>Yes</u> ☐ <u>No</u>
6. read each question carefully?	☐ <u>Yes</u> ☐ <u>No</u>	☐ <u>Yes</u> ☐ <u>No</u>	☐ <u>Yes</u> ☐ <u>No</u>	☐ <u>Yes</u> ☐ <u>No</u>	☐ <u>Yes</u> ☐ <u>No</u>
7. avoid messy calculations and work out longer calculations <u>on a scratch sheet?</u>	☐ <u>Yes</u> ☐ <u>No</u>	☐ <u>Yes</u> ☐ <u>No</u>	☐ <u>Yes</u> ☐ <u>No</u>	☐ <u>Yes</u> ☐ <u>No</u>	☐ <u>Yes</u> ☐ <u>No</u>
8. avoid wasting time on answering too much, making <u>unnecessary calculation</u>s?	☐ <u>Yes</u> ☐ <u>No</u>	☐ <u>Yes</u> ☐ <u>No</u>	☐ <u>Yes</u> ☐ <u>No</u>	☐ <u>Yes</u> ☐ <u>No</u>	☐ <u>Yes</u> ☐ <u>No</u>

9. pace yourself so as not to spend too much time on any single problem?   ☐ <u>Yes</u> ☐ <u>No</u>   ☐ <u>Yes</u> ☐ <u>No</u>   ☐ <u>Yes</u> ☐ <u>No</u>   ☐ <u>Yes</u> ☐ <u>No</u>   ☐ <u>Yes</u> ☐ <u>No</u>

10. watch out for little errors that could become big ones, sign mistakes, sloppy arithmetic?   ☐ <u>Yes</u> ☐ <u>No</u>   ☐ <u>Yes</u> ☐ <u>No</u>   ☐ <u>Yes</u> ☐ <u>No</u>   ☐ <u>Yes</u> ☐ <u>No</u>   ☐ <u>Yes</u> ☐ <u>No</u>

11. write clearly and neatly?   ☐ <u>Yes</u> ☐ <u>No</u>   ☐ <u>Yes</u> ☐ <u>No</u>   ☐ <u>Yes</u> ☐ <u>No</u>   ☐ <u>Yes</u> ☐ <u>No</u>   ☐ <u>Yes</u> ☐ <u>No</u>

12. check your answers to see if they make sense?   ☐ <u>Yes</u> ☐ <u>No</u>   ☐ <u>Yes</u> ☐ <u>No</u>   ☐ <u>Yes</u> ☐ <u>No</u>   ☐ <u>Yes</u> ☐ <u>No</u>   ☐ <u>Yes</u> ☐ <u>No</u>

Relax

A little nervousness is natural. Too much nervousness will obstruct your thinking process and handicap your mind. The first thing to do is to relax. Take a few slow, deep breaths. You will find that the increased oxygen in the brain will do wonders for your concentration.

How to take an exam

Learn how to take an exam. There are several important points to keep in mind:

1. Read the directions for the exam, if there are any.

2. When you first get the exam, assess the amount of time you will need in order to complete the exam. If there are 20 questions and you have one hour to complete the exam, then you have less than three minutes per question. If some questions are given a heavier weighted score than others, give those questions proportionately more time.

3. Scan the entire exam for familiar questions. First try the questions that look easy.

4. Read each question carefully. Begin writing your solution only after thinking about alternative (perhaps simpler) solutions.

5. Do not waste time by giving results that are not asked for. If the answer is $\sqrt{2}$, do not calculate it to be approximately 1.4142136.

Very important!

6. Check your answers. Are they reasonable? Do they make sense?

While taking the exam, you should think about:

1. Are you answering what is being asked? Take time (seconds) to understand exactly what is being asked.

2. Are you making careless mistakes? (Slow down, if you are.) Most careless mistakes are in the arithmetic and algebra. Watch out for sign mistakes and mistakes with parentheses. If you have time at the end, you should check your results before you hand in your paper. You will be more relaxed at the end of the exam and more likely to catch your careless errors at that point.

VERY IMPORTANT!

3. Are you running out of time? It is important to pace yourself at all times. **Do not spend too much time on one question!** Spending a disproportionate amount of time on one question is likely to be your biggest worry. Avoid doing it at all cost.

4. RELAX! (Easy to say; hard to do.) It's only an exam. You are not performing brain surgery.

Tips

- Read the directions, if there are any
- Compute the average time you should spend on each problem
- Scan the exam and do the easier questions first
- Read each question carefully before writing a solution
- Do not waste time by solving more than you are asked to solve
- Check your answers for careless mistakes
- Be sure your answers are reasonable
- DO NOT SPEND TOO MUCH TIME ON ONE QUESTION
- RELAX

Summary 9

Of critical importance is how much confidence you have in your ability to solve problems in calculus. Every suggestion in this book is designed to build confidence. If you are organized in space and time, you will feel good about pursuing your studies, because you will work more efficiently.

If you carefully listen in class and take good notes, you will have a clearer understanding of what to study and of what your are expected to know.

If you read the text properly and work through the assigned problems, using the techniques suggested, you will have more success at answering more problems correctly.

If you prepare your assignments as suggested, your assignment grade will increase and that alone will help build your confidence.

Never forget quiz preparation. Never put yourself in a position where you have to cram for a quiz or exam.

Ask your instructor for help whenever you think you need it.

Learning calculus takes time and effort. There is no quick way. There are no guarantees to success at learning any mathematics, but organization and the guidelines of this book will help. Keep two things in mind as you go through the course:

1. You learn mathematics by doing it-not by reading it, or watching it done
2. Do the best you can. Be patient with it and understand that it does take lots of work and time.

Answers to Diagnostic Evaluation

	Answers	Area of weakness
1.	$\dfrac{x+3}{x+2}$, $x \ne -2$	exponents, fractions and/or the function concept
2.	$h = \dfrac{108 - x^2}{4x}$, $x \ne 0$.	isolating unknown qualities in an equation
3a.	$x = \pm 4$;	
b.	$x = -1$.	solving quadratic equations
4a.	$x = 0$ and $x = 2$	exponents and logarithms
b.	$x = \dfrac{1 + \log 13}{2 \log 2}$ or $x \approx 2.35$	
5.	$A = 64$ units.	geometry of circles; circumference of circle
6.	$x = 1$	quadratic formula, radical equations ($\dfrac{1}{4}$ is an extaneous root)
7.	$x = \dfrac{-14 + \sqrt{456}}{2}$ and $y = 22 - \sqrt{456}$ or $x = \dfrac{-14 - \sqrt{456}}{2}$ and $y = 22 + \sqrt{456}$.	
		simultaneous equations and quadratic equation
8.	$x = 2$ and $y = 3$	simultaneous equations
9.		graphing polynomial functions

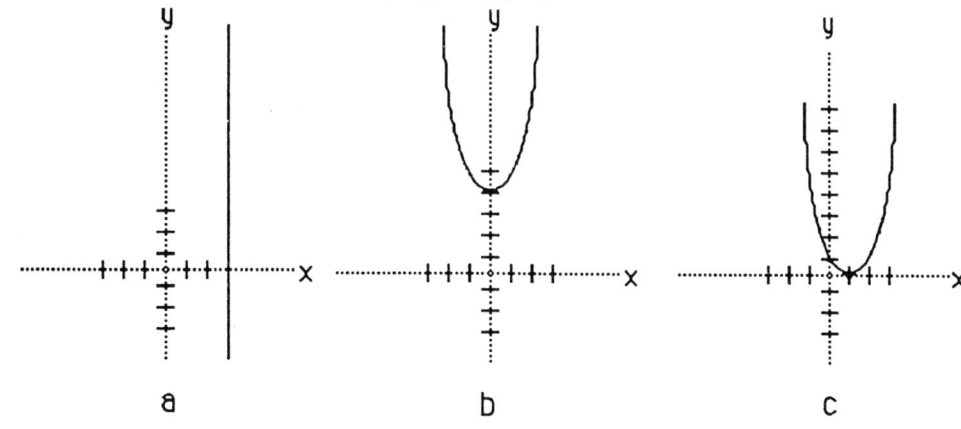

a

b

c

Answers to Diagnostic Evaluation

10.   graphing polynomial functions

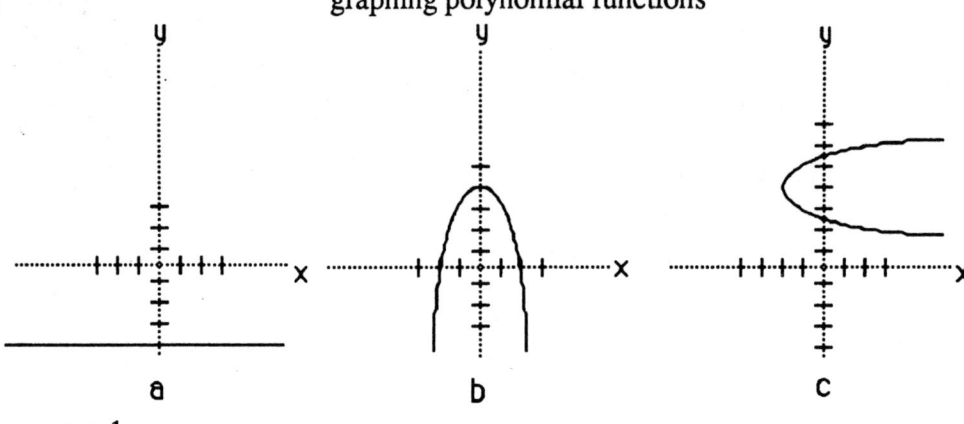

      a   b   c

11. $\dfrac{x+1}{x}$, $x \neq 0$.   working with compound ratios

12. $x = 3\sqrt{3}$ and $z = 6$.   trigonometry / Pythagorean Theorem